THAR

The
Great Indian
DESERT

T H . A R

The Great Indian DESERT

R C Sharma

Lustre Press
·
Roli Books

ISBN: 81-7436-057-3

© **Roli Books Pvt Ltd 1998**
Lustre Press Pvt Ltd
M 75 Greater Kailash II Market
New Delhi 110 048, INDIA
Tel.: (011) 6442271, 6462782
Fax: (011) 6467185

Text: R C Sharma
Wildlife section and captions: Kishore Singh
Map: A Z Ranjit

Photo credits
Amit Pasricha, Hashmat Singh, Joanna Van
Gruisen, J L Nou, Karoki Lewis, Rajpal Singh
Subhash Bhargava, Lustre Press Library
Fotomedia
Aditya Patankar, Akhil Bakhshi
Amrit P Singh, Ashish Khokar
Sanjay K Saxena, Shalini Saran

Conceived and designed at
Roli CAD Centre

Printed and bound at
Star Standard Industries Pte Ltd, Singapore

Acknowledgement
The author wishes to express his gratitude to
Mr Vijai Verma, IAS, for giving permission to use
his personal library and Prof H S Sharma for
all logistic support.

Contents

Previous pages 6-7:
*The twilight hour in the
desert flames different
shades of orange, a time
when the camels and
cattle are brought back
home while a velvety
darkness spreads quietly
across the desert.*

Pages 8-9:
*The Rabaris
are a colourful desert
community whose love
for colour and
adornment of the body
with tattoos is a
prominent characteristic.*

Pages 10-11:
*Visitors
may be fazed by the
savage living conditions
of the Thar, but to the
Rajput it is a land his
forefathers wrested from
the wild with
tremendous courage,
and he will not be
parted from it. His
companion, quite
naturally, is the camel,
a beast that has adapted
remarkably well to the
living conditions of this
dry countryside.*

*Nomad Banjara dancers
in Jaisalmer. Nomads
prefer to travel rather
than settle, and are
the ancestors of the
community of gypsies
worldwide.*

Following pages 14-15:
Jaisalmer's Sonar Qila or golden fort once stood like a sentinel on the ancient trading routes that connected countries in Central Asia with India and beyond. Though the fort is still inhabited, newer houses have breached its battlements to populate the desert beyond it.

Pages 16-17: *The royal families have given up their titles, but still participate in full regalia on important state occasions and festivals. Here, the former Maharaja of Jodhpur, Gaj Singh, wears the traditional tie-and-dye* safa *or turban and carries a sword. With him are his mother, wife and son.*

Pages 18-19: *The wedding of Ram and Sita, a scene from the* Ramayana, *painted on a* haveli *in Mahensar. The Shekhavati region is a scrub desert with small towns to which the Marwari community of businessmen trace their origin. Their wealth was used to build lavish homes that were profusely painted, for all practical purposes turning the area into an open-air art gallery.*

SHIMMERING SANDS AND COLOURFUL CLANS

The mighty Thar Desert has been immortalised in songs and legends as *Maroosthali* the land of the unknown—hostile, harsh and merciless. There is, however, another side to this story.

To anyone who has visited the great Thar, it conjures up images of a blaze of brilliant colours on the turbans of men and the *ghagras* (ankle-length skirts) of women. It is the land of grand beards and bristling moustaches, the land where the common folk are fiercely proud of their history—with good reason too. The Thar is the lap which cradled the ancient Indus Valley civilisation and gave rise to its well-organised settlements, rich harvests and progressive culture.

Further down the ages, there were the sumptuous temples of Osian, Ranakpur and Ramdeora, the colourful fairs of Nagaur, Kanana and Marwar, the frescoes of Shekhavati—like an open-air art gallery; the richly engraved and adorned gates and balconies of Jaisalmer which are living architectural wonders, the fortified palaces of Mehrangarh in Jodhpur and Junagarh in Bikaner, and even later, the royal salons of the Palace on Wheels train, the life-giving waters of the Indira Gandhi canal, and the conversion into heritage hotels of historic castles such as Khimsar and Mandawa; all these attest to the cultural and social achievements of this desolate and arid region.

It is so easy to slip into the glorious era of the daredevil Rajputs when you are sitting on a vast dune listening to the mellifluous voice of Chimme Khan, the old Manganiyar singer, narrating the sagas of valorous Rajput kings and their equally courageous queens.

The royal history of the Thar is a fascinating tale of Rajput clans and dynasties vying with each other to settle permanently in the sandy wastes of the Marwar and Godawar plains on the western slopes of the Aravallis. Castles and

Above: *The elaborate rituals of Rajput weddings turn them into truly princely affairs. The bride is veiled and the unveiling ceremony is an important part of the celebrations.*
Opposite page: *If the sands of the Thar replicate the sea-bed, it is perhaps because of a distant memory they carry of a time, millions of years ago, when it was originally covered by a vast sea. According to an Indian legend, the sea dried up when Lord Rama fired an arrow at it, creating immense heat and vapourising the water.*

forts, magnificently constructed and proudly conserved, by the Bhattis of Jaisalmer and the Rathore Rajputs of Marwar and Bikaner are living testimonials of this long and fierce struggle.

Down the centuries, the sands of the Thar have also provided refuge to defeated kings like Humayun, rebel princes such as Khurram (later Shah Jahan), and lesser mortals like outlaws, vagabonds and gypsies.

But before that story, let's go back to the beginnings of the great Indian desert. Surprisingly, there are no positive paleo-botanical evidences about the circumstances which led to it. History suggests that it must have come into existence sometime between the end of the Pliocene and the last glaciation periods, when wide-ranging climatic and weather fluctuations could have given rise to such a difficult environment. The region became increasingly arid towards the end of the Eocene times, which could imply that sometime between the early and the mid-quaternary period—well after the advent of humans—it turned into a desert. Archaeology, especially Indus Valley findings, suggests that desert conditions were accentuated only in the last 5,000 years.

The arid conditions and precarious nature of rain-fed agriculture are mentioned in Kautilya's *Arthashastra*, written around the third century BC. From Ashoka's times (240 BC) to the times of Skanda Gupta (457 AD), agriculture here was only possible with the help of irrigation. Huien-Tsang, the famous seventh century Chinese traveller, had to cross a formidable stretch of desert to reach Sindh from Gurjardesh (Barmer). An archaeological study in 1952 concluded that 'the Rajputana (Thar) desert is at least two millennia old and its nucleus must be older still'.

The region was completely devoid of any human settlements in the pre-historic period, shunned by the paleolithic and neolithic savages. Even in the proto-historic period, there are no traces of human habitation; except in the northern parts where the Saraswati Valley had been a commingling of many rivers, not just geographically but also culturally. Archaeologists, after extensive studies of various remains, came to the conclusion that this valley was in the dried-up beds of Ghaggar in Ganganagar district. These early settlements continued to exist until the seventh and eighth centuries, after being revived by the Greyware and Rang Mahal cultures. The beginning of desiccation and desert environment became quite evident by the first century BC.

The invasions of the northern plains by the Huns of Central Asia led to large-scale immigration, resulting in a great influx of people. These refugees settled in the comparatively more hospitable regions, leaving the dreary and desolate parts of the Thar and the Magra.

Gurjaras, the ancestors of the modern Gujars, were one such group. They established themselves in the beginning of the proto-Rajput period and enriched the pastoral economy. In their struggle for survival, the Central Asians and Gurjaras came into bitter conflict and this battle for supremacy started the process of building large, walled and fortified settlements.

Opposite page: The Rajputs, their money-lenders and prime ministers, the Jains, as well as the traders, the Marwaris, propitiated their gods and built several temples to mark their devotion. The pacifist Jain religion co-existed with the more martial tradition of the warrior Rajputs with no apparent conflict.

After the Gurjaras were ousted, a new phase of settlement and survival started. New kingdoms and princely orders came up like that of Mandore, near Jodhpur. The northern border of the region was under separate rule and called Janglo Desha, based at Nagaur. Viratanagra (Shekhavati) was another hub for concentration of power of the Matsya kingdom. The Bhatti clan took charge of the western reaches of the Thar and established a kingdom at Tanot; this was later shifted to Lodurva, near Jaisalmer.

The later medieval or Rajput era was, and remains, a trendsetter in many ways. The essence of a Thar settlement was established during this period. Royal palaces perched on high ground and fortified urban centres were the norm of the day in the late fifteenth century throughout the Thar—Jaisalmer, Jodhpur, Bikaner and on to the Shekhavati region.

The Chauhans at Nadol (Pali), the Rathores at Jodhpur and Bikaner and the Parmars and Solankis in Shekhavati were the leading dynastic houses of the time. Practically all the Rajputs clans and dynasties came to the Thar as fugitives to re-establish themselves in strategically safer places. Even years later, feuds among the Rajputs resulted in new branches of dynasties. War was the religion of the Rajputs, and they went to the battlefield to slay or be slain.

The most daredevil among them were the Rathores of Marwar, who were known for their audacious valour; a fact that earned them many enemies. So much so that the armies of Marwar—known as the *panchas huzaar turwar Rathoran* or the fifty thousand swords of the Rathores—were always kept in readiness to rally around the *gaddi* or throne of Rao Jodha, the ruler of Marwar.

An interesting tale is told in the context of one of Rao Jodha's campaigns in Mandore, the erstwhile capital of the Rathores from where the invader Muhammad Ghori had driven them out in 1211 AD: it is said that a bard in Mandore had revealed before the battle that, 'the bird of omen is perched on Rao Jodha's lance and the star which indicated his birth shines bright on it'. This apparently signified that the Rao would be sure to succeed in his campaign. By a curious coincidence, the king did capture Mandore, and went on to re-establish the glory and supremacy of the Rathores all over Marwar.

However, having exhausted outsiders to war with, the Rathores turned upon themselves, as court intrigue and inter-sibling rivalry ran rife in the handsome palaces and *zenanas* (ladies' wings) of Marwar. The outcome was a split in the family. Rao Bikaji, the sixth son of Rao Jodha, founded his independent kingdom in 1488 AD, which came to be called Bikaner. All this fighting, however, kept their armies in excellent shape and the Rathore cavalry continued to be regarded as the best, right up to the first half of the British period.

Meanwhile, royal towns became symbols of personal and princely pride—a manifestation of stately order, affluence and architectural excellence. Artisans and craftsmen found patrons and many small-scale industries began to flourish.

Ironically, development and security went hand in hand with devastation and insecurity. The royalty, for all its grandeur, brought with it internecine wars and jealousies resulting in periodic destruction and forced migration. This was further intensified on account of the Muslim invasions in these centuries.

Rajput royalty became virtually subservient to the Mughals in the fifteenth and sixteenth centuries. Throughout Mughal history the region was under the *subah* (province) of Ajmer with *sarkans* (sub-divisions) at Jodhpur, Bikaner and Nagaur and the *subah* of Delhi with its Rewari *sarkans*. Later the Pindaris, Pathans and Marathas wreaked disaster here and everyday life became so miserable that it resulted in another large-scale exodus.

By 1818 AD, the suzerainty of the British was firmly established in the Thar, with its headquarters in Ajmer. After a long gap, an era of peace and prosperity was ushered into this desolate region. New road and railway networks were laid down, modern irrigation systems helped revive agriculture, recent mining innovations were introduced and modern industries, along with newly-planned market places (*mandis*) became the order of the day.

Describing the market town of Pali, Colonel James Todd wrote in his book, *Annals & Antiquities of Rajasthan* (1877): 'Pali was the entrepot for the eastern and western regions, where the production of India, Cashmere, and China, were interchanged for those of Europe, Africa, Persia and Arabia. Caravans from the ports of Kutch and Gujarat followed the route via Sooie Bah, Sanchore, Beenmahl, Jhalore to Pali. The most desperate outlaw seldom dared to commit any outrages on caravans under the safeguard of these men (*Charuns*), the bards of Rajputs...'

However, there were difficult times on account of severe droughts and famines. The eighties and nineties of the nineteenth century were extremely challenging and exacting on account of intense and recurring famine conditions. Entwined with the glorious tales of Rajput kings and striding progress encouraged by the British, is the unending saga of misery, tyranny and calamities faced by the common people in face of nature's onslaught of famines, droughts and occasional floods. The capricious nature—often made worse by a self-indulging, quick to take to the sword royalty—forced the common man to uproot himself every other year; a constant reminder that he was helpless and subservient to forces superior to him.

SHIFTING DUNES AND FLOWERING SHRUBS

The word Thar symbolises a desolate sandy tract. It carpets 2,08,751 square kilometers, an area equal to sixty-one per cent of the state of Rajasthan and around seven per cent of India. Transcending the Indo-Pak boundaries, it spreads through twelve districts of Rajasthan and some parts of northwest Gujarat, especially in the peninsular Kathiawad and the Rann of Kutch. Gujarat, or more precisely Kathiawad, is the region from where sand has blown incessantly over thousands of years to the Thar. But it is just the source region and not the nucleus of the desert. The core of the desert lies in parts of Jaisalmer, Phalodi, Pachbhadra, Bikaner and Barmer. The eastern boundaries of Shekhavati form its extension, merging with the semi-arid to sub-humid Jamuna plains of western Uttar Pradesh.

The Thar is populated by 17.51 million people, which is forty per cent of Rajasthan's and two per cent of India's total population. The spatial density varies widely—from nine persons per square kilometer in Jaisalmer to two hundred and fifty-three persons per square kilometer in Shekhavati (1991 Census).

Physiographically, the important units of the Thar are the Ghaggar plains in the north, the upland region alongwith the Aravalli out-crops on the north eastern margins, the Godawar tracts on the western slopes of the Aravallis, the Magra rock-floored plains scored by windblown sand in the midwest and the area fully blanketed by sand dunes in the west. Between the sand ridges of the Great Thar are low-lying areas which act as ephemeral drainage basins during sudden downpours in the rainy periods. This intermediate depression, known as *marho*, is rich in alluvium—a reason why most villages are concentrated here. Areas in the western part of Sam, Ramgarh and Nachna sub-divisions have no drainage and form an absolute dreary sea of thick sand dunes, matching the desolateness of the Magra rocky wastes.

Above: *The camel is ideally suited to the landscape, existing on the leaves of the* khejri *tree and surviving on water stored in its body for up to a few days.*
Opposite page: *There is immense beauty in the harsh landscape of the desert, and even the camels, blackbuck, bustards, cranes and peacocks that find the habitat conducive, have a somewhat awkward elegance.*

Mirages of magnificent ponds, lakes and seas, made so romantic by the *Arabian Nights,* are a common sight here. Incidentally, mirages are actually short-lived visual impressions caused by the sun's searing rays falling on the arid and sandy landscape.

The northern part is called the Little Desert, on account of the nature and magnitude of dunes, which are mostly transverse in form and smaller than their counterparts in the Great Thar. There are minor depressions between the dunes called *gassis,* which during the rains become shallow water basins and are important for the survival of the villages.

For vast stretches, the desert is parched, its only source of water—the rains—often playing truant. Continuous years of drought are not unknown.

A major battle for survival in the Thar is against sand and sand dune formations. Sand blown by wind piled up in heaps and ridges or spread in sheets, abounds in various parts of the Thar. These dunes cover more than 40 per cent of Jaisalmer, 47.8 per cent of Bikaner, 36 per cent of Barmer and 53 per cent of Churu districts. Most of the sand movement is due to saltation which also decides the type of sand dune. Other factors which make dunes differ in their formation are suspension and surface creep. The quantity of sand in this movement depends on the wind velocity above and the nature of suspension underneath it.

There are two systems of sand dunes—old and new. The old system was formed by intensive dune building during the earlier prolonged arid climatic phase. These are well-stabilised, vegetated and cemented. The new ones are still in an evolutionary phase. They are active, smaller in size and have distinct morphological characteristics. The new dunes are a great menace to agricultural fields, settlements, lines of communication and irrigation systems.

Sand dunes have a distinct ecology and their flora go through various stages of development and degradation. Therefore, while navigating through sand dunes and encroaching sand, one has to be careful not to interfere with

Salt pans are another feature of the Thar. The areas around Sambhar Lake, in particular, are used for the making of salt.

their ecological succession which is of crucial importance in the greening of dunes.

The local people have developed some devices to cope with the creeping sand, but unfortunately they are not very effective in the absence of vegetational barriers. The road from Churu to Sardarshahr in the Shekhavati region, and from Jodhpur to Pokaran, is often blocked by creeping sand dunes, making vehicular traffic impossible for several hours. Complete settlements are buried under the burgeoning sand; the Paliwal settlements located on the highway leading from Jaisalmer to Barmer amply testify to this example of nature overpowering man.

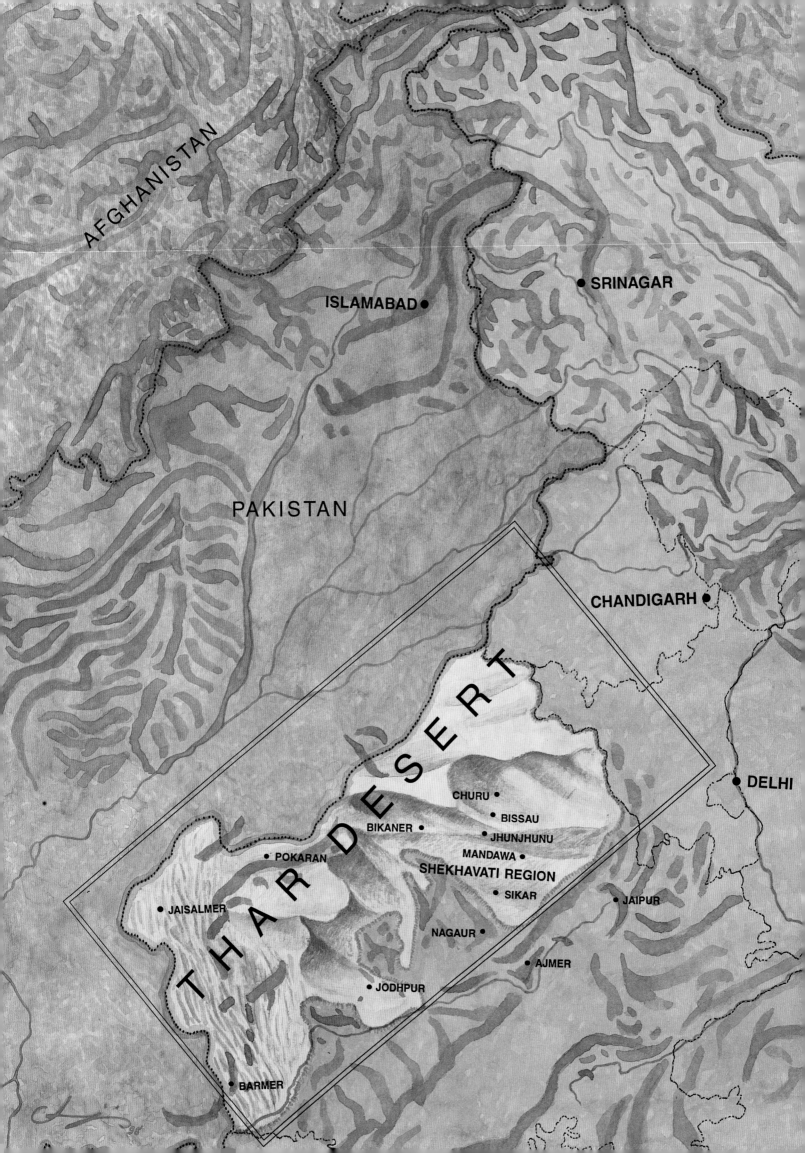

CITIES, PALACES, FORTS AND TEMPLES

Ajmer: Said to have been founded in 145 AD, this pilgrimage town is holy to the Muslims.

Places to Visit: The **Dargah of Khwaja Muinuddin Chishti**, the tomb of a twelfth century Sufi saint, is the second holiest site after Mecca. There is also the **Mayo College**, the **Ana Sagar,** an artificial lake, and the **Nasiyam Jain temple**. The town of **Pushkar** is a day's excursion from here.

Barmer: In the heart of the desert, this town is a centre for wood carving, carpets and embroidery.

Places to Visit: The **Kiradu** temples built in the eleventh century are known for their intricately sculpted columns.

Bikaner: Founded by the Rathore chief Rao Bhikaji, Bikaner was the centre of the cross-desert caravan trade.

Places to Visit: The imposing **Junagarh fort** (1588-93) is one of the best preserved in Rajasthan. The palaces within are beautifully designed with superb carvings, fine lattice work and lacquered doors. The red sandstone **Lallgarh palace** has some of the best carvings and *jali* work. **Bhand Sagar** (5 km from Bikaner)—a group of sixteenth century Jain temples—and the **Karni Mata temple** at Deshnoke (33 km) as is a day's excursion from Bikaner.

Jaisalmer: This twelfth century city still retains its medieval charm with its narrow lanes, historic forts and famed *havelis* or multiple houses within an enclosed, surrounding wall.

Places to Visit: The **Jaisalmer fort**, famous as the golden fort due to its colour, stands atop the Trikuta Hill. Another must-see is the early nineteenth century **Patwon ki haveli** which has the finest collection of murals and carved pillars. The **Garisar**, a little water body, attracts numerous migratory birds and has small shrines which are worth a look.

Jodhpur: This second largest city of Rajasthan was once the capital of the Rathores.

Places to Visit: The majestic **Mehrangarh fort** dominates the city as it rises from a steep hill. The museum within has an unparalleled collection of weapons, jewellery and musical instruments.

The **Umaid Bhawan Palace** built by Maharaja Umaid Singh is the pride of Jodhpur. The 347-room palace is strikingly English with its indoor swimming pool, private auditorium, eight kitchens and a ballroom. **Jaswant Thara** is the cremation ground of the former rulers of Jodhpur. Built in white marble these commemorative structures house portraits of successive rulers of Marwar.

Nagaur: About 137 km from Jodhpur is the **Nagaur fort**. It's a must-see for its wall paintings, temples and *havelis*. It also has a shrine of the disciple of the great Sufi saint Muinuddin Chishti.

Shekhavati Region: This loosely held confederation of towns is known for its profusely painted *havelis*. Some of the best can be seen in the towns of Mandawa, Samode, Ramgarh, Lachhamangarh and Bissau. **Osian** is home to the largest collection of eighth to tenth century Jain temples.

During the fifties and sixties the march of the desert was a topic of heated debate and worry. The vigorous creeping of the Thar towards the north and northeast, making the land infertile, has been recognised as an ongoing disaster and steps have been initiated to introduce large-scale planting of trees in Punjab, Haryana and western Uttar Pradesh under the *Van Mahotsav* (afforestation) programme. Recently, active schemes have been initiated to stabilise the dunes and green them by planting *sewar*, a local grass rich in nutrients.

Another grim battle for survival in the desert is against the extreme climate of the region. The Thar is undoubtedly the hottest area in India with very low

There is something incredibly romantic about the desert, and despite the unbelievable loneliness, it continues to haunt the imagination, stirring the senses just as the wind teases the sand.

relative humidity. Maximum temperatures during the day often reach 41°C to 48°C and one of the major tasks of man is to protect himself from the sun and the heat. But December to February is markedly cold as the region comes under the influence of the western disturbances and the temperatures often touch freezing point.

However, there are micro-level factors which affect the climate of each region. The weather, for instance, becomes extremely sultry in the southwestern parts when the whole region comes under the influence of high pressure during the monsoon. Similarly, cold and wet conditions prevail when the northern parts pass under the northwest depression and freezing temperatures are common here.

The climatic conditions that can range from severe droughts to devastating floods, are crucial to living in the Thar. Of particular importance are fluctuations in rainfall. Chronological studies reveal the dismal fact that a general famine is expected once in ten years and a drought once in four years.

The period from 1894 to 1904 was marked by severe drought conditions, resulting in unforeseen miseries and hardships. The years 1932 to 1939 again saw harsh drought conditions. A particularly dreadful drought occurred in 1987

The essentially pastoral communities of the Thar survive on farming and cattle. Villagers prefer the strong-smelling, sweet goat milk, and the hardy creature is able to survive well, often travelling long distances for a nibble on scrub bushes.

affecting the lives of almost 25 million people and 30 million livestock, prompting an unprecedented exodus of man and cattle.

Droughts reduce the level of water, make it brackish, lower the stock-carrying capacity of pasture and reduce crop production. Fodder shortage often leads to butchery in different parts of the desert and villages shrink as people

Overleaf: *Communities in the Thar live mostly in scattered villages. Though huts have given way to pucca houses, the farming methods are still primitive, and electricity still a luxury.*

abandon their homes and farms and flee to the neighbouring tracts of Malwa in Madhya Pradesh and the Jamuna plains in Uttar Pradesh.

The other feature of the Thar is the hot sand and dust laden winds called *Loo* which blow enormous quantities of sand on the roads and the railway tracks, making movement difficult. Like the *Khamsin* of Egypt, *Simon* of Saudi Arabia, *Shammal* of Mesopotamia and *Brick-Fielder* of Australia, the *Loo* is very trying and dreadful during the afternoons. At times it continues even into the night.

The attire, abode and food of the Rajasthanis help them cope with the onslaught of the *Loo*. Clothing accessories like turbans, *odhinis* to cover the

A typical Thar village consists of mud-walled homes with thatched roofs that offer refuge from the loo, *the hot summer wind. A fence of thorny branches keeps out wild animals.*

head and *angrakhis* (tunics) reduce the impact of the heat and the blowing sand. The hutments have a typical form such as round-shaped *jhonpas* surrounded by thorny, dried-up bushes and walls across the openings and gateways. The Rajasthani breakfast consists of cooling herbs like mint, lemon and so on. A typical meal would feature butter-milk, red chilli chutney and raw onions—all of which help in dealing with the *Loo*. The local people follow two

Facing page: *These women gathered at the village well are wearing* chooras *or their wedding bangles. Earlier of graded ivory, these are now more usually made of bone or even plastic.*

basic rules of survival: never leave home on an empty stomach, and carry lots of water when you set out.

The vegetation cover is scanty to begin with and gradually shrinks and disappears as one goes from the slopes of the Aravallis and the adjoining Godawar tract to the sandy wastes of the Great Thar and the rocky barren surface of the Magra. Among the prominent plant species are *babool, khejari, aak,* cacti shrubs and occasionally *sesum, neem, peepul, jamun* and prickly *phok.* New varieties of trees like eucalyptus are grown abundantly. Other flora include thorny plants like *Anogeissus pendula, Acacia benegal, Acacia catechu* and *Boswellia sonata* with shrubs like *Salvadora oleoides* and *Maytenus emarinatus.*

The xerophytic vegetation which prevails, rather survives, over the vast tract of the Thar does not provide shade and relief from the scorching sun. Over-grazing by camels, sheep and goats and cutting by man further devastates even this meagre greenery. The Shekhavati region provides a glimpse of this rampage: naked and stunted trees with almost no foliage stand on roadsides. Such scenes are common in other parts of the region too, the only exceptions being the Gang Canal and Indira Gandhi Canal command area where pleasant green vistas of plantations refresh the eye.

No big game can be seen here but there is distinct desert fauna adding colour to the aridness. The Thar does not provide a suitable habitat for wildlife such as the tiger, panther and sloth bear but is home to desert fox, jackal, sambhar, porcupine, wild boar, chinkara, desert cat and blackbuck. Out of the two dozen national parks and wildlife sanctuaries in the state of Rajasthan, only three sanctuaries are located in this region.

The National Fossil Park near Sam in Jaisalmer covers an area of about 3,162 square kilometers whereas Tal Chhapar in the district of Churu has only eight square kilometers. The Desert National Park is the homeland of the great Indian bustard locally known as *godawar.* Once a common bird, the *godawar* is now an endangered species found only in the Desert National Park. The safe confines will hopefully give the bird a new lease of life. Tal Chhapar has a unique ecological setting with a natural depression and abundant salt formation in the neighbourhood. Around 1,400-1,500 blackbuck inhabit this site and move in large herds.

The Wildlife Protection Amendment Act (1991) provides enough guidelines for the preservation of rare species facing partial or full extinction. Sanctuaries elsewhere in the state like Keoladeo Bird Sanctuary (Bharatpur) and Ranthambhor Tiger Reserve (Sawai Madhopur) are ecologically rich and have become popular on the tourist calendar of anyone visiting the great Indian desert. Of the thirty-two natural enclosed areas—restricted areas meant for preservation, where hunting is prohibited—in the country, Rajasthan accounts for nineteen.

The question of inter-dependence between man and flora and fauna assumes prime importance in an environment where nature is not benevolent. Meagre resources, inclement weather and a fragile eco-system conspire to set limits to human interaction with nature. For a visitor the dazzling sunsets over the golden sands may seem mesmerising, but for the Rajasthani it merely represents relief from the blistering heat of the day.

Wildlife in the Desert

The desert provides a fascinating habitat for wildlife, and though the naked eye may at first be fooled into believing that there is little to be found in the arid landscape, the Thar, in fact, supports an incredible variety of flora and fauna. But then, there is remarkably little 'pure' desert in Rajasthan, much of it being scrub, with acacia and short, thorny trees being the rule rather than the exception. These trees and shrubs

Blackbuck stage a mock fight: this beautiful deer is particular to Rajasthan and besides being extremely agile, is also one of the most graceful of the species. A single buck usually has the run of the harem, while the rest of the males are confined to the fringes of the herd.

provide protection, however inadequate, for a variety of wildlife ranging from mammals to reptiles and birds.

Ecologically, the vegetation in much of the Thar comes under the 'thorn forest' category. Prolonged and intense biotic interference, however, has taken its toll, transforming the natural cover. Cultivation is the principal culprit in this exercise, and even in the remaining area, the density of the natural cover is much less than what the soil and climatic conditions permit.

The shrubs and trees that cover the dunes include phog (*Calligonum polygonoides*), kair (*Capparis decidua*), oak (*Calotropis procera*), kumata (*Acacia senegal*) and khejri

(*Prosopis cineraria*), while rohida (*Tecomela undulata*), peelu (*Salvadora oleoides*) and ber (*Zizyphus nummularia*) occupy their base. The sandy plains have scattered tree growth along with low shrubs and grasses such as sania (*Crotolaria burhia*), kheenp (*Leptadenia pyrotrehnica*), bui (*Aerva javaraca*) and sewan (*Lasiurus sindicus*). To reduce the loss of water by evaporation, the plants tend to reduce the size of the leaves or even do away with them altogether.

Since there is a shortage of water in the desert, among the predators one does not find the tiger and leopard, otherwise widespread through the state. However, it is not unusual to find the desert cat, or even the desert fox with its bushy tail. But there is a surprisingly large population of deer, especially blackbuck and chinkara. Other mammalian species include the desert hare and wolf.

The Thar has a high avain diversity of over three hundred birds species, but a low avain endemicity due to effective physical or ecological barriers restricting the movement of such species. Most of the birds found in the region have a wide distribution. Migratory birds are known to winter at the water spots in the desert, whether natural or man-made (specially for *shikar*). The imperial sandgrouse is one of the most popular species spotted here, as is grey partridge, both game birds for the table. Birds of prey, on the other hand, include the tawny, short-toed as well as spotted eagle, the lagger falcon and the kestrel. The great Indian bustard is offered special protection, since it was once close to extinction. This large bird lives in small flocks, and feeds on cereals, dung beetles, grasshoppers, locusts, lizards, small snakes and berries. Other important species include the houbara, demoiselle crane, pintail sandgrouse, imperial sandgrouse, white winged tit, king vulture, cinereous vulture, peafowl and the cream courser. A large number of other birds are to be seen too, among them cranes, bee-eaters, drongos, orioles, shrikes, larks, geese and teal.

Top to bottom: *Saw-scaled viper, long-eared hedgehog, blackbuck, desert fox.*

The desert is also home to a number of reptilian species, including the spiny tailed lizard, desert monitors (resembling miniature dragons), sandfish, gerbils and chameleons. While the landscape supports a number of snakes, the deadliest of the poisonous ones include the Sind krait, Russel's viper and the saw-scaled viper. The sand boa and large rat snake are quite common.

Most species have adapted special mechanisms to survive the scarcity of water. For one, they reduce physical activity considerably and avoid exposure to the hot winds, hiding in the shade of shrubs or by burrowing deep into the sand. The earth is an admirable insulator and an animal even a few centimeters below the surface will comfortably survive the hottest day or the coldest night. In the Thar, from foxes to cats, lizards and snakes, all enter their burrows to avoid the hot, desiccating air outside. Most of these animals are active in the early morning, or after sunset, when the temperatures are much lower. In the case of animals such as the chinkara, a rise in body temperature up to seven degrees above normal can be tolerated without serious injury to any vital organ. Chinkaras, for example, can go without water for a number of days by feeding on the aak plant to absorb water from its leaves. Similarly, the desert fox and cat hunt gerbils for their high moisture content.

Two of the principal national parks and sanctuaries in the Rajasthan Thar include the National Desert Park and the Tal Chappar Sanctuary. While the former was set up to provide protection to the great Indian bustard, the latter is known for its large population of blackbuck.

A unique feature of the desert is the community of Bishnois who live in the Thar and do not allow either the felling of trees or the killing of animals. Spread out between Jodhpur and Bikaner, this pastoral community fiercely protests all killings in its lands, which is why it is not unusual to see herds of deer grazing openly around Bishnoi villages, seeking the shelter of trees only to escape from the heat.

Top to bottom: *Great Indian bustard, demoiselle crane, Indian sandgrouse, black ibis.*

SMALL SETTLEMENTS AND
SPRAWLING CITIES

Ak ra jhopra, Phok ra bar, Bajra ri rooti,
Mot'h ri dal, Dekho ho Raja, tharo Marwar.
(Huts of aak; barriers of thorns; breads of maize;
lentils of the vetch; behold Oh! King; your Marwar)

This is a common local saying in Marwar and one can find no better description of the austere lifestyle that people of this region lead, and yet it also speaks of their spirit—which takes even these hardships with a pinch of salt. Indeed, the people of the desert might be weather beaten, but they have never lost their zest for a colourful life; in fact, the regional folk culture has only been enriched with the passage of time.

The extreme weather and frequent seasonal fluctuations are a challenge that humans face afresh every morning. Even the daily changes are strong enough to test his power of adjustment and potent enough to completely demobilise him.

This ability to innovate has, incidentally, helped the Rajasthanis elsewhere. The ingenious people of the Thar make excellent businessmen and this region has sent out thousands to the great metropolises of India and abroad. In fact, among the top twenty business houses of India, the majority are from the Thar and its sub-regions—the Birlas, Dalmias, Bangars and Singhanias all belong to the Marwar or Shekhavati region. Frugality and innovativeness have been imparted to them almost as part of an ethnic and race memory and the ability to adjust to any emerging situation—be it famines, droughts or occasional floods—seems to have helped them become extremely dynamic and hence, successful businessmen.

Despite the challenges that the desert environment offers, people have settled all over the Thar and have innovated in their own small ways to make the arid

Above: The men of the desert are easily distinguished by their flowing beards and colourful turbans. The manner in which a turban is tied as well as the way a beard is worn is enough to tell one about the region and even the community to which they belong.
Facing page: A family of Rabaris with their distinguishing dark coloured skirts and veils, and wealth of silver jewellery. Even children wear a large amount of jewellery about their bodies.

sands habitable. There are agricultural and pastoral settlements; villages that have become pilgrimage centres; there are settlements along the river bank or wherever water is to be found, fortified shelters offer sanctuary, while jobs are to be found in mining towns and at seasonal fairs or *melas*.

An analysis of the rural landscape helps in understanding the role of various factors and processes of settlement. For the very large villages, the landscape is an outcome of the complex factors of location, culture, caste, community, nature of the economy and so on. Next to them are large villages which do not contain a well-established village market and have a vague street pattern. Medium size villages are conspicuous by a medley of houses and tortuous lanes, no shops or street pattern, and the houses are generally *kutcha* (temporary). Small villages are typical of arid lands and are more in harmony with the natural environment: thatched houses and huts (*jhonpas*) huddle in a most irregular way with no street pattern or definite layout.

The central place is occupied by either a village well or a temple as in the case of the village Mukam where all social and cultural life revolves around the temple of Jambheswarji founded in 1593 on the *samadhi* (grave) of the saint. Water is, of course, the deciding factor in their location, except in the case of villages like Goriya which are situated on the Aravalli tract where water is plentiful.

The communal canvas of rural Thar presents a mixed picture. The Rajputs, Jats, Bhambhis, Bishnois and others dot this caste-ridden landscape. In some villages, certain communities dominate, as in Malar which is a Pushkarna Brahmin stronghold. An interesting tale is told about the Pushkarnas who, so it is believed, trace their origin to the builder caste. They apparently pleased the gods by building the sacred lake of Pushkar and, hence, were upgraded as Brahmins (the highest caste in Indian society). Till today, the chief object of icon worship among them is the *Kodali* or the pick-axe used for digging. Incidentally, the Pushkarna women are noted all over the Thar for their beauty and grace. There is a saying in Rajasthan which goes:

Marwar is the mine of robust men,
Jaisalmer of beautiful women,
Sindh of excellent horses
and Bikaner of camels.

This might have been written with Malar in mind—since the cultures of the three erstwhile states of Jaisalmer, Jodhpur and Bikaner are mirrored here.

A very good example of the Rajasthani sense of humour in the face of adversities, is to be found in village Kalijal, forty kilometers from the city of Jodhpur. Tradition has it that the village was founded by one Hariya Jal. The villagers, perhaps in gratitude or hope (since *hariya* means greenery), named it Hariya Jal. However, it failed completely to live up to the name and was a constant victim of droughts and famines. The villagers were so disgusted that they changed its name to Kalijal, which means black fire!

However, many of the villages are so old and remote that their origins and history can be determined only by the inscriptions on the memorial stones of Rajput *satis*. *Satis* were women who were burnt on the funeral pyre of their husbands. At one time, this was quite a common practice in Rajasthan.

The most colourful villages in the Thar are to be found on the Shekhavati tract. These have well-built houses, more often then not with painted walls and beautiful decorations and wall paintings.

Habitations in the Thar present a common picture: the settlements are choked because of the all-pervading sand dunes, except for those on the slopes of the Aravalli and the richly cultivated Godawar region, where the Luni with its numerous tributaries has changed the landscape. In fact, the capricious weather and highly erratic rainfall often results in large-scale movement of cattle and herdsmen.

The cattle breeders have well set routes that they follow year after year. They remain in the nearby areas of Gujarat, Madhya Pradesh, Punjab, Haryana and parts of eastern Rajasthan for months, even years, if the difficult weather continues. Sheep and cattle migration routes go northwards from Jaisalmer and Nagaur to Ganganagar across the Indira Gandhi Canal area; eastwards from Balotra across Jodhpur to the eastern tracts of Ajmer and the Harouti area of Bundi, Kota and Jhalawar, and from Jodhpur and Pali south towards to Gujarat across Jalore and Sirohi. The camel migration routes are two directional—east-west and north-south, that is Jaisalmer-Ganganer and Jaisalmer-Nagaur, Jaisalmer-Jalore, Barmer-Ajmer, Bikaner-Churu-Sikar and Alwar.

The meagre fuel resources of the regions has resulted in the virtual loot of nature. The scanty vegetative cover of the Thar is being mercilessly cut for food and fodder. It has now reached a level where not even roots are being spared. A common sight in Shekhavati and parts of Jodhpur and Pali, are men and women carrying bundles of wood, unmindful of the disaster they are causing.

Ironically, Rajasthan is also the perfect example of man's ingenuity and tenacity in preserving bio-diversity, not because of laws or the dictates of religion, but out of the simple desire to conserve the biotic resources of their land. A unique example of this comes from Jodhpur in Marwar where one can watch herds of blackbuck and Indian gazelle roaming in the fields and surviving on the desert vegetation, totally unconcerned about the people around them. The Bishnoi villages of Guda Bishnoian, Doli and Dhawa located near Jodhpur city firmly believe in preserving trees and animals and protect them against armed poachers. Khejerli is another village with the same avowed principles and practices of protection and conservation. The blackbuck, antelope and the *khejri* tree are considered sacred by the villagers and for their protection, they are willing to even risk being shot and killed by poachers. There is a well-documented episode that occurred way back in 1737 when Gridhar Das Bhandari, an officer of the erstwhile Jodhpur state ordered the felling of *khejri* trees. Despite the official sanction, the strong and stubborn villagers would not allow even a single tree to be struck down.

Since then, this has acquired the status of a Khejerli Movement, like the Chipko Movement of the Uttar Pradesh-Garhwal area in northern India. The commitment of the entire community towards these norms and rules is unique

Overleaf: A sprawling village on the outskirts of Jaisalmer: this could be a biblical city, so remote is it from modern civilisation. Unlike other cultures in north India, the people of the Thar seem to be in no hurry to catch up with the rest of the world.

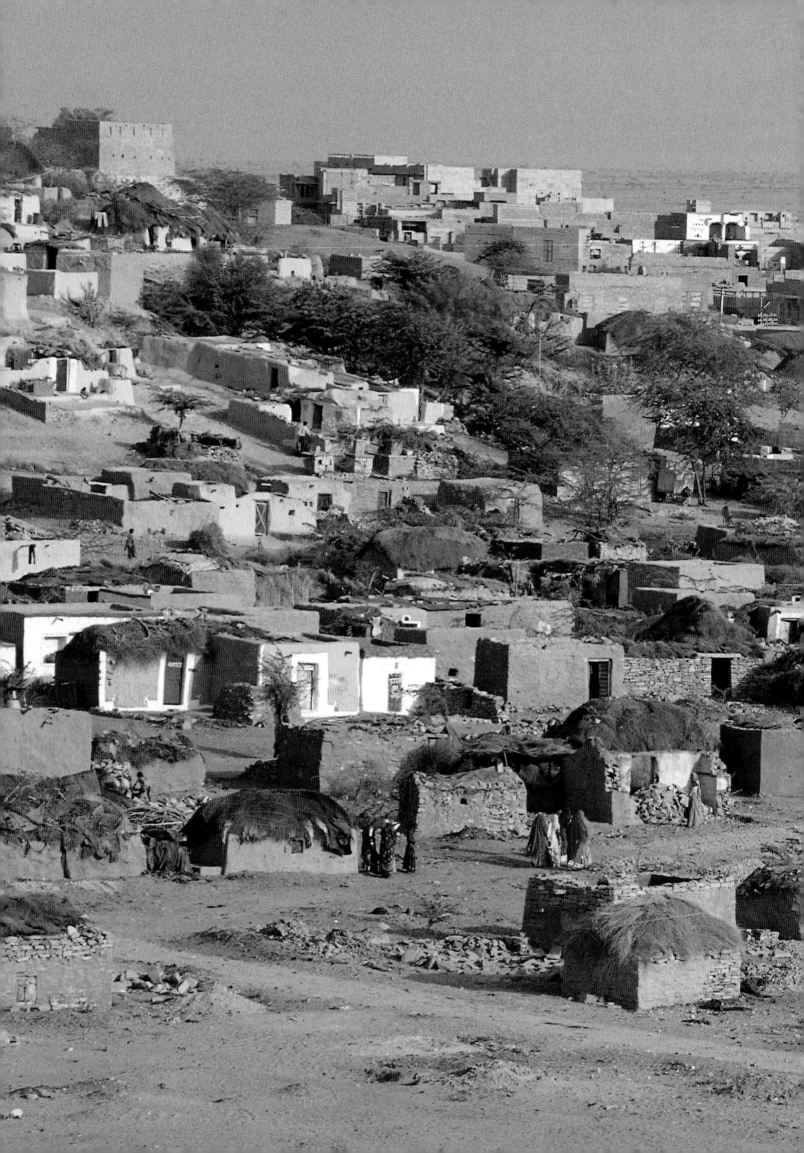

and they stand apart from countless other communities, not only of the Thar but throughout India.

If the villages of the Thar are dotted with *jhonpas,* the cities feature a variety of architectural forms and structures. They depict either varying forms of adjustment with the inclement weather or intense love and pride for architectural richness and extravagance. Some of the towns show excellent town-planning and settlement development. Although habitations are designed keeping in mind the climate, they are also products of the political and cultural history of the region.

The typical jhonpa *or village hut is covered with a layer of mud and cow dung which is then painted over in bright colours to lend a little colour to the landscape.*

The fabric of the settlements are heavily dotted with non-urban nuclei in accordance with the general Indian pattern. In fact, Bikaner and Ganganagar are essentially overgrown villages. Ever since their inception, they have acted as the focus of trade and commerce and the sites of rendezvous for caravans from all directions of the region. The imperial cities of Jodhpur, Bikaner and Jaisalmer have been the seats of great political authority and grandeur.

Some self-sufficient rural villages persist even today and a compact settlement with its tank or well and a struggling bunch of acacias, tamarix and zizyphus in the midst of yellowish sand is still the dominant feature of the landscape. Just as water is the *raison d' etre* for the location of villages, truly urban centres and cities are often associated with a fort perched on a hill, a palace surrounded by a haphazard collection of houses and enclosed by a city wall, the market occupying the central position on the roads joining the opposite gates.

The waves of invasions since the tenth century and the consequent devastation and onslaught have had their effect on the morphology and size of the settlements. Defence has played an important role in this respect. Rural habitations on account of this overriding factor are big in size and well nucleated in structure. Out of the 14,356 rural settlements (1991), some are very

The desert communities are self-sufficient, even spinning their own blankets using camel hair. Their lives have an austerity that is impossible to imagine at the threshold of the new millennium.

large and compact with a population of more than ten thousand people. The region has seventy per cent of Rajasthan's large villages indicative of security and water requirements as the stabilising factors.

The late medieval or the Rajput period (1200-1857) laid the basic fabric of the settlement, while the feudal system consolidated its stakes and permanently settled

Overleaf: *Camels and cattle are led to water sources by villagers at specific hours. Sometimes these are set away from the main well while, at others, well water spills are trained to run into shallow troughs where the water is gathered for the purpose.*

its boundaries. However, this was destabilised in the eighteenth and nineteenth centuries on account of the Pindaris and the Marathas. The British brought in an era of peace and tranquillity to usher in proper development of settlements. A new set of administrative towns were developed to properly knit together the rural habitations. Railways, roads, irrigation, increased mining activities and industrial development along with trade and commerce focused on newly developed *mandi* (market) centres which helped in the growth and development of settlements.

It is interesting to note that the names of the settlements indicate the importance of natural and cultural factors. Many names indicate local features like *garh* (fort): Ratan*garh*, Hanuman*garh*; *mer* (rocky surface) for Jaisal*mer* and Bar*mer*; or availability of water like *sar*, *sir* and *tala* like Naga*sar*, Rawat*sar*, Bhim*sir* and Buran-ka-*tala*; or factors of ethnicity like *Bishnoi*-ki-dhani; *Jaton*-ki-dhani and *Pali* named after Paliwal Brahmins; specific functions like *ner*, *shahr*, *nagar*: Bika*ner*, Sardar*shahr*, and Ganga*nagar*; and rural centres with *bas*, *was*, *khurd* and *dhani*: Charan*was*, Jogra*was*, Nawa*bas*, Raniwara-*khurd* and Hira-ki-*dhani*.

The urban centres present an interesting amalgam of natural, social, cultural and historical factors. The word *garh* (castle) as a suffix gives evidence of their origin and importance. Further, urban morphology and structure betray the effects of the chequered growth of these centres. The city's population and functions provide a basis for urban classification and typology. Some of the distinct types are agricultural towns (Pipar and Bilara), mining towns (Khetri), industrial towns (Ganganagar), commercial towns (Phalodi and Nokla Mandi), transport towns (Sojat), administrative towns (Nawa), service towns (Jhunjhunu), educational towns (Pilani) and diversified towns (Barmer and Nagaur); in addition are the two cities, Jodhpur and Bikaner. All together there are eighty-four urban centres of varying sizes and nature—Shekhavati is decidedly the most urbanised with thirty-three urban centres. Next comes the Marwar tract with twenty-six urban centres.

The city forms or morphological structures are results of the movement of people and their establishments. Centrifugal and centripetal forces act in association to decide the layout of a city. In the case of the Thar, these have acted in a chequered manner in response to the historical and political contours of the region. Perhaps, the best way to describe the manner in which these cities and urban centres have evolved is to select specific cases.

Jodhpur

Rao Jodha, in 1459, founded the capital city of the Rathores of the erstwhile state of Marwar. Located in a depression, the city seems to have all that is needed for an urban culture. It is surrounded by yellow stone hills which have provided its building material. The Umaid Sagar and Hemawas schemes supply adequate water.

The old fort on the isolated elevated rocky surface and the royal palace are all within fortified walls. The four exit gates are the most prominent features of

Opposite page: Mud-painted walls receive a sudden, effusive burst of colour with simple motifs and patterns. The interiors of these huts tend to be cool, while pucca houses tend to absorb heat and make the interiors stifling.

the city. The city walls were expanded in the eighteenth century and two more gates added. Later additions beyond the walls include a railway station, new residential colonies and other such amenities. Jodhpur has been a trading centre along with a strong defence and administrative set-up. Within the walls, in the narrow lanes are market places like Girdikot, Kotla Bazaar and Subzi Mandi. Industrial establishments like tie-dye, calico printing, *badla* (water pot) making and shoe and ornament making came up later. In fact, even an industrial estate came into existence in 1950. The new city administrative quarters, High Court and university are later developments.

The royal families, with the wealth of the kingdoms over which they held sway, built impressive fortified cities such as the Mehrangarh fort at Jodhpur. The marble memorial, Jaswant Thara, stands in the foreground.

Jodhpur is a sprawling urban centre with a population of about half a million and has a major role to play in the state. That said, however, it has its medieval and princely charm and architectural splendour in the shape of forts, palaces and temples, most using the yellow sandstone popular as a building material, and with very little use of marble.

The people of Jodhpur are celebrated for their courteous behaviour and cultured way of living. Whether you ask for water, or the time, or a great favour, the answer will come laced with a soft-spoken, '*Jee huzoor, hukam*'

(Your wish, Sir, is my command). Since the sixteenth century, Jodhpur has been known for its leather work, spices, *jutis* (light and brightly decorated camel-skin sandals) and *mawa kachoris* (sweet filled with thickened milk).

The city land use and functional segregation in the form of residential, industrial, commercial, open-space and institutional areas stand witness to the excellent planning of the city. The city has grown along with the people who have built it over the years. It retains much of its medieval temperament which has not been swept away or obscured by modern developments. It continues to be the prime city of Marwar, and of the Thar.

Mehrangarh, typical of desert forts, chose a high location on which to raise its battlements, a defensive strategy, though the palaces located in the highest apartments have delicate windows and balconies since these would have been difficult to scale.

Bikaner

Founded in 1485 by Rao Bika, Bikaner is a typical Thar city with its sand dunes, prominent brick-coloured buildings and the Lallgarh palace. Being on a slightly raised ground, the forts, palaces and spires of temples tower imposingly

Overleaf: Under the eyrie-like battlements of Mehrangarh fort, the city of Jodhpur sprawls in what looks like a complicated maze. The white-painted houses are built along narrow lanes to offer maximum protection from the overhead sun.

as symbols of stately elegance and grandeur. Junagarh fort around which the city has grown, was founded by Maharaja Rai Singh.

In the twentieth century, there has been a slow growth in the north, northwest and eastern directions. Among the recent landmarks are Lallgarh palace in the north, the railway workshop in the northwest, the Victoria Memorial, the Residency and so on. However, most of the modern residential areas have come up on the main arteries joining the city. The city did not enjoy rapid and vigorous growth as economic opportunities within the city have been limited in the absence of industrial development.

Within Bikaner's Junagarh fort, the royal band still gathers to play a stirring, martial beat from the past. Such fortified settlements consisted not only of palaces but also all the defense and administrative offices concerned with the running of the princely states.

Geographical determinants have virtually shaped the morphology of this urban centre. Cultural factors in the form of numerous temples, mosques and Jain monasteries have also been a major influence. Old market areas such as Sarat Bazaar, Ghas (grass) Mandi (market) and Vaidyaon ka chowk (doctor's square) are confined within the walls or adjacent to the city gates. Among recent additions outside the old settlements are Station Road, Fard Bazaar, Dhan (grain) Mandi and Katla Bazaar which came into existence with the city's expansion.

Being the seat of power of the erstwhile princely state, the city has been prominent in promoting culture, industry, trade and commerce. It has also provided residential complexes to the aristocracy.

The cottage industries are wool bailing presses, metal works, oil mills, *dal* (lentils) and flour mills, shoe-making, carpet-making and so on. Bikaner is also famous for its salted delicacies, like *bhujia*, and huge variety of sweets. The city covers more than fifty square kilometers of area and has been divided into distinct land use zones like commercial, industrial and residential. There are areas earmarked for the economically weaker sections like *sadhs, bambis,*

A moat around Junagarh offered protection and large gates guarded the entrance. Within the battlements were apartments built in lattice-like tiers to capture any hint of breeze.

chamars and *mehtars* and such areas suffer on account of unhygienic conditions. The Oswals, Jaiswals and Maheshwaris belong to the middle-class residential section and have better living conditions. The economically prosperous communities have elaborate housing complexes with all public amenities. The lower, middle and upper class residential areas lend the city a mixed personality, where the social and economic factors override the dictates of nature. Bikaner presents a mixture of indigenous and western morphology and yet it retains a medieval character. Of late, water from the Indira Gandhi Canal has stepped up growth and development.

Osian

Osian, a desert town situated sixty-three kilometers north of Jodhpur, is almost a pilgrimage centre for lovers of art and culture. It is an ancient township and was earlier called Upa Kosapur. Through the years, it has remained the most important pilgrimage for Hindu devotees of the Vaishnava, Saura and Sakta sects. It contains the largest group of eight to tenth century Jain and Hindu temples in Rajasthan.

The typical Osian Pratihara temple is set on a terrace whose walls are finely

The incredible opulence of the durbar *halls in Bikaner was the result of artistry, combining the love of painting with gold to achieve the effect of inlay. Though the walls are not marble, the gloss could certainly lead most people to assume so.*

decorated with mouldings and miniatures. The sanctuary walls have central projections with carved panels and above these rise curved towers. The doorways are usually decorated with river goddesses, serpents and scrollwork. The twenty-three temples are grouped in several sites north, west and south of the town. The southern group of small Hindu temples includes one to Harihara (Shiva and Vishnu of the Indian trinity). The west group contains a mixture of Hindu temples, including the Surya or Sun temple (early eighth century) with beautifully carved pillars. The Jain Mahavira temple (eighth to tenth century), on a hillock, is the best preserved of the temples here. The Sachiya Mata temple (eleventh to twelfth century) is a temple to the goddess Durga.

Jaisalmer

Jaisalmer is the capital of the erstwhile princely state founded in 1156 by Rawal Jaisal of the Bhatti clan. It is a fort town with basations around it for security. Magnificent rock engravings on yellow stone are found abundantly here. Besides the royal palaces and forts, even the houses of the Jain traders and ministers

Close to Jodhpur is Osian where a large number of temples pay tribute to the gods. The residents of the desert, so exposed to the vagaries of a harsh nature, continue to revere the elements which they believe control the pattern of their lives.

show richly ornamented balconies and facades. Patwon ki Haveli is a classic and magnificent example of this.

The paucity of water and natural vegetation has today made it a struggling town with a desolate and dreary landscape. Caste-based residential areas still exist within the fortified city for Rajputs, Pushkarna Brahmins, Maheshwari Jains,

Overleaf: *A caravan of camels winds its way through the desert. Motorised transport has changed the mode of carrying goods, but camels are still used to pull carts, carry water or even humans across sands where no roads exist.*

Kumars and other communities. The main commercial areas are near the fort on the Phalodi road. Stone-carving, woollen blankets, handloom and clarified butter or ghee-making are its small-scale household industries.

The landscape around Jaisalmer is particularly desolate, and it is therefore all the more impressive to see a city of such magnificence having established itself in the desert. What Jaisalmer did not have—nature's bounty—was more than adequately compensated through the wealth it commanded on the trade routes of Asia, making it possible to hire master-artisans to craft the splendid dwellings of Jaisalmer.

Cenotaphs flank a shallow depression of water on the edge of Jaisalmer. Known as Garsisar, such memorials were often raised to kings and important persons in medieval desert society.

Ganganagar

This city is a product of the new agricultural development. With the advent of the Gang Canal in 1927, this dry region was transformed into fertile, thriving

Facing page: The facade of Patwon ki Haveli. This haveli started the grand era of mansions in Jaisalmer. These sandstone buildings were most excessively carved, with lavish balconies, jharokhas *or windows, awnings, pillars, balustrades and eaves, yet the result was harmonious to the eye.*

land that attracted settlers from the nearby areas. It is a typical canal settlement with planned residential, commercial and industrial areas. The economic base is strong and the migrants, specially those from the neighbouring plains of Punjab, have made this settlement dynamic and vibrant. The urban landscape and ethos are foreign to the traditional Rajasthani cultural ambience. Unlike other cities Ganganagar is an urban culture, the product of abundant water, planned efforts and Punjabi values. It lacks the imperial elegance and magnificent medieval structures usually associated with urban centres in the desert.

Castle Mandawa in the Shekhavati region consists of a fortified settlement built in a complex maze of inter-connected buildings. Though a part of it is a heritage hotel today, the aura is still fiercely medieval, and from the parapets it is easy to see a stretch of the scrub desert that surrounds it.

Shekhavati

The townships in Shekhavati are different from most others in the region. They are not planned in the true sense but are elegant on account of their superb frescoes or wall paintings. Shekhavati is a highly urbanised area and is dotted with dozens of urban centres, well-connected by metalled roads. The sprawl of the settlements with their special characteristics have succeeded in mellowing the effects of the harsh environment with extremes of temperature and plenty of sand dunes.

Among the Shekhavati townships, the district towns of note are Churu, Sikar, Jhunjhunu and some heritage townships like Mandawa. Commerce and trade are a legacy of the past when some of these townships were an important caravan nuclei. Otherwise, most of them are either overgrown rural settlements or headquarters of *tehsils* or *thikanas*. The morphology remains ill-developed and confused on account of the physical and social contours.

The painted interiors of Shahpura Palace have retained the turn-of-the-century flavour that marked the palaces. While the rhythm is distinctly Indian, the pace accommodated innumerable western mores.

The love for architectural elegance and creativity is evident both in the case of rural and urban habitations, in various ways. The Thar has a princely legacy and articulate religious beliefs. The royal families built palaces, forts and cenotaphs, whereas the trading community encouraged the construction of temples and places of worship. Both are illustrative of the love for architectural extravagance and minuteness of form, colour and content.

SUMPTUOUS ARCHITECTURE AND ENDURING CULTURE

Royal palaces are proud reminders of the Thar's colourful past and occur in somewhat embarrassing abundance. They reflect the local and regional traditions as much as the perceptions and whims of the rulers and the impact of alien styles. Each region and sub-region boasts of structures that defy all environmental impediments. The Thar is dotted with archaeological sites with remains of forts, palaces and mansions like Tilwara (Barmer), Kalibangan and Rang Mahal (Ganganagar), Sunari (Jhunjhunu), Ganeshwar (Sikar), Mandore (Jodhpur), Lodurva (Jaisalmer), Bhinmal (Jalor) and Nadol (Pali)—dating from different eras of its historic past.

Among the more famous contemporary palaces are the Umaid Bhawan and Ajit Bhawan palaces in Jodhpur and the Lallgarh Palace in Bikaner. Mandore, the Rathore capital before Jodhpur, is characterised by the range of its architecture. The Ekthamba Mahal was built for special social functions and Zenana Mahal for the use of the royal ladies; they are amongst the finest palaces in Rajasthan. Along with the complex is the Maharaja Ajit Singh (1671-1724) Museum which preserves local sculptures and architectural wealth. It also contains the remarkable miniature paintings on the *Raga Raganis*. Dozens of statues of gods and goddesses and inscriptions are stored here. Ajit Bhawan Palace, the residence of Maharaja Ajit Singh, has been turned into a heritage resort. The palatial mansions nestling in lush green rustic settings are unique. Each bungalow is named according to the month of the Indian calendar and its zodiac sign.

Umaid Bhawan Palace is located on the outskirts and is perched on a high mound, giving it a brooding air. The panoramic view from up there—you can see the entire countryside from the ramparts of the gigantic dome—is alone worth going all the way up for. It is perhaps the largest personal residence in

Above: *The former royals have traded their heritage for commerce. Maharani Usha Kumari, the consort of Maharaja Swaroop Singh of Jodhpur, for example, runs Ajit Bhawan as a hotel.*
Facing page: *Maharaja Dalip Singh and his wife at the ancestral castle which they have just renovated. The aristocracy consisted of feudal chiefs who owed allegiance to their princes and supplied them the armies to expand and consolidate their kingdoms.*

the world. Made of stone, it has murals by a Polish artist and baths carved out of single blocks of marble. The huge underground swimming pool takes one's breath away. Large personal collections of art and artifacts speak of the elegant taste of the royals. A part of this palace has been converted into a heritage hotel too.

Lallgarh Palace in Bikaner is located slightly away from the city and is considered one of the finest buildings of Rajasthan. Designed by Swinton Jacob, it has exquisite carvings that decorate various nooks and corners of the palace. They depict an interesting and intriguing union of European comforts and oriental fantasy. Among the different wings are Laxmi Vilas and Shiv Vilas. Most of the palace is now part of a heritage hotel, while seperate parts house a museum and the residence of the erstwhile royal family.

Apart from the royal palaces are the forts and castles scattered throughout the length and breadth of the Thar. Some are massive while others are small—depending upon the status of the erstwhile states. Important *thikanas* have magnificent castles, many of which have been converted into heritage hotels and guest houses. The fort in medieval times served as the nucleus for the establishment of settlements and was symbolic of security and administration. Among these edifices, the most impressive are the Mehrangarh fort of Jodhpur, Junagarh fort, Bikaner, and those at Nagaur and Jaisalmer.

Jaisalmer fort presents a beautiful silhouette against the setting sun. Cobbled pathways ending in a succession of gates—Akhaspol, Ganeshpol and Hawapol—lead up to the lofty fort. This massive and impressive fort was built in 1155 by Rawal Jaisal. It is made of yellow stone and perhaps this is why it came to be known as Sonar Garh (fort of gold). Like most other forts in this region, Jaisalmer fort has witnessed several instances of Rajput valour. A typical example is the resistance that a chieftain, Rawal Dudda Trilokshi and his army of brave Bhatti warriors offered the Sultan of Delhi, Firoz Shah Tuglaq (1351-1388) when the latter attacked Jaisalmer. It is said that all of them perished fighting. The women were not lacking in valour and rather than submit to humiliation at the hands of the conqueror, committed *johar*—a Rajput tradition of self-immolation practised by the ladies when the men were defeated by a conqueror—ending their life by jumping into a communal fire.

Inside the fort is the breathtaking Sheesh Mahal, the royal palace built by Maharaja Aakh Singh. The grand Moti Mahal made by Mool Raj and the exquisite Gaj Vilas and Jawahar Vilas palaces are also part of the complex. The fort has Jain temples of the eleventh and twelfth centuries which are excellent examples of art and engraving. It also houses rare religious manuscripts.

Jalore fort near Jodhpur is believed to have been built by Nagbhatta I in the eighth century. However, in the absence of proper care, it is now in a dilapidated state.

Jodhpur's Mehrangarh is a symbol of the Rathores. It was established in 1459 by Rao Jodha and is located on the imposing hill Chidatoonk, about six kilometers away from Mandore. It has the appearance of an elegant peacock and so is also known as Moradwaj Garh. In the fort are magnificent palaces and buildings like Moti Mahal, Phool Mahal, Rani Bas, Top Khana, Vichala

Mahal and the Sringar Chauki where the kings were crowned. The Mehrangarh library houses a rich collection of rare books and manuscripts.

The castle of Siwana, about thirty kilometers from Jalore, is another magnificent fort belonging to the tenth century. Nagaur fort, built by Amar Singh Rathore, is one of the most strategically placed forts and has witnessed much bloodshed and bitter battles to occupy it in the past. Mohammad Ghauri, Qutub-ud-din Aibak and Maharana Kumbha are among those who have held this fort at various periods.

Bikaner's Junagarh fort was believed to have been impregnable. Constructed by Rai Singh around 1589-1594, it has a moat surrounding it. As a matter of fact, the fort never did see much battle since the ruling family enjoyed friendly relations with the Mughals.

Its gates—Karampol, Chandpol, Daulatpol, Ratanpol and Dhruvpol—are exquisite pieces of work. The fort has beautiful palaces and a grand royal *chowk* (square). Successive kings added buildings like the Janani Dhyadi, Gaj Mahal, Phool Mahal, Lal Niwas, Ganga Niwas, Anup Mahal and so on.

Other fort towns in the Thar include Nawalgarh, Lachhmangarh, Ratangarh and Nagaur and follow the princely tradition. Besides these imperial establishments are other monumental *chattris* (cenotaphs), temples and *havelis*. *Havelis* are generally town houses, as opposed to *kothis* or garden houses. The word *haveli* is of Persian origin and means an enclosed place where people live together with common consent; they are usually inhabited by joint families. Secluded from the outside world, a *haveli* sets its own pace of life. All through royal and feudal India the *havelis*, whether inhabited by Hindus or Muslims, represented the rigid lifestyle of a society that segregated its men from its women. The typical *haveli* in Shekhavati consists of two courtyards—an outer and an inner. The grander ones sometimes have three or four courtyards. The architecture reflected the socio-cultural position of women who normally remained in *purdah* (veiled) and retreated to the *zenana*, their private apartments, when the men were around. Today, life in the *havelis* continues on much the same pattern, though there are fewer inhabitants and the inhibitions too are relaxed.

Cenotaphs are pieces of art, erected in the memory of princes, kings and elders. They abound in the whole region but particularly so in the tracts of Shekhavati and Marwar. Jaswant Thara, a nineteenth century cenotaph, is located near the fort of Jodhpur. Built of white marble, it presents a beautiful contrast to the sandstone hill it stands on.

Mandore has four elegant cenotaphs of earlier Rathore princes. A curious aside is the cenotaph of Maratha general Appaji near Nagaur which stands as a reminder of the frequent attempts by the Marathas to conquer this area.

Havelis built by kings, princes, noblemen and rich businessmen abound in

Following pages 72-73: The Maharaja of Jodhpur still resides in the imposing Umaid Bhawan Palace built in the art-deco style and completed in the 1940s. Fortunately, it is so huge that while part of it is a museum, a large chunk also operates as a hotel.
Pages 74-75: The sumptuous Anup Mahal in Bikaner's Junagarh fort is lavishly decorated. The throne where the maharajas sat during formal durbar is richly canopied. This is one of the rare palaces where red has been used so excessively—and dramatically.

various parts of the Thar. However, the Shekhavati tract is proverbially known for these *havelis:* delicately ornamented and decorated by frescoes. So much so that Shekhavati is considered to be an open-air art gallery.

After the decline of the Rajput nobility, the prosperous Marwari business class became the patrons of fresco art. Brisk trade and commerce heralded prosperity and the Shekhavati regions of Churu, Sikar and Jhunjhunu are littered with these big mansions. All the lime-plastered walls of these *havelis,* temples, rest houses (*dharamsalas*), barns (*gaushalas*) and water-wells are profusely covered with frescoes.

The havelis *of Shekhavati are profusely painted, often depicting scenes from everyday life, such as rulers, warriors and ceremonial processions. These have now become a record of a society at its most important point of transition.*

Traditionally wall paintings started as purely religious in form and theme, but slowly they moved towards ornamentation and imitation of European lithographs and etchings. The early ones used natural pigments. Colours were made into a paste, and when required, applied on the wall surface while still wet. This is known as the *fresco buono* technique and combines a sheen with durability. The secret lies in the smooth plastering of the walls for then they can withstand the environment for centuries.

The arrival of synthetic dyes changed the *fresco buono* style to *fresco secco* style, which is painting on a dry surface. The workers are highly skilled artists trained under a master artist and are known as *chiteras* or *chejaras* (masons). Payment to them depended on the area, style and content of the fresco.

The contents of the frescoes vary according to the perception, wish and competence of the artists as well as the budget of the patrons. There are simple themes portraying peacocks, parrots, deer, elephants, tigers, lions, horses, camels, mango groves and peepul trees with creepers. The sun and the moon are also used profusely on the plastered walls.

Subjects like folk heroes and battlefields along with religious themes

Carved doorways and lintels were surrounded with paintings ranging from the secular to the religious. Here, an obviously important woman has travelled especially to confer with a guru *or religious teacher.*

dominate. The *chhatri* ceiling or the whole front wall on both sides of the main entrance often vividly depict epic scenes from the *Ramayana* and the *Mahabharata*. Folk tales like *Dhola Maru* and *Krishna Leela* with richly coloured and ornamented *gopis* are frequent subjects of these wall paintings. *Krishna Leelas* are based on the life and loves of the Hindu god Krishna. He is said to have danced with all his *gopis* (lovers) simultaneously on full moon nights. The story has various erotic and esoteric shades and is a popular theme of paintings all over India. *Dhola Maru* is a love narrative enacted by local professionals in many parts of the Thar to the accompaniment of music. A couplet says, 'Come quickly Dhola. It is the third day of *savan* (rainy month of July) and your beloved Maru will die of fear seeing the lightening.'

Western culture and British rule have left an impact on the content and form of these fresco paintings. *Sahebs* and *memsahebs* replaced some indigenous themes. Today even cars, palanquins, aeroplanes, acrobatic shows, wrestling scenes, everyday chores, portraits of ancestors, saints and holy men are common themes. There are unusual scenes of women shooting arrows or flying kites. The painters had a free hand and were encouraged to let their imagination rule the form and content of the paintings.

Some of these fresco-decorated *havelis* and mansions are well-known and frequently visited by tourists. The most celebrated is the Sone Chandi ki Haveli (gold and silver mansion). Built in 1846 by Seth Ram Poddar at Mahensar, it has finely painted frescoes amidst bewildering opulence. Sukhdeo Das Ganeriwala Haveli built in 1880, with its erotic paintings and religious scenes of Radha and Krishna, is another outstanding example.

The unusually blue-coloured Biyani Haveli in Sikar delights visitors and is charmingly reminiscent of English porcelain. The obsession with blue is understandable: synthetic blue began to be imported from Germany around 1870. For a wealthy patron, using blue must have been a way of establishing primacy.

The Thar is abundantly dotted with sumptuous and lavish temples dating from 700 onwards. Before that, this area had suffered from intermittent raids and hence could not support large places of worship. The Mandore and Osian temples date back to the early Pratihara period of 700. In Osian, on the raised topography, the temple complex starts with Satyanarayan and Sun temples. The Mahabir and Harihar temples stand out as important landmarks. The temples have decorated terraces and graceful, single-turreted spires. *Mandaps* (halls) are a distinct part of these temple complexes and some of them have pyramidal roofs such as that of the Mahavira temple.

The middle phase temples are represented by the Kameshwara temple at Auwa and the Ranchhodji temple at Khed near Balotra. Here are seen the beginnings of the multi-turreted arrangement over the sanctum.

The Harsha temple southeast of Sikar, Kekind south of Merta, Paranagar near the Sariska wildlife sanctuary and Kiradu west of Barmer reflect the denouncement of the Gujara-Pratihara style. The Neelkantheshvara temple at Kekind has excellently balanced embellishments on the outer walls of the sanctum. Each side of the temple accommodates as many as thirteen images. The *mandap* has beautifully carved pillars.

At Kiradu, interestingly located in a natural hilly amphitheatre, is a group of five temples with multi-banded plinth and multi-turreted spires. The *mandap*, although not in perfect shape, combines strength and elegance. There are some surviving *toran*-arches of the Vishnu temple.

Kiradu symbolised the onset of a new ornate style known as Solanki. The Chaumukha Jain temple near Sadri (Pali) is a remarkable Jain temple in sylvan surroundings. It has four openings with two distinct *mandaps* that make the temple appear magnificent and grand. Beautiful ceilings, airy interiors and dizzying heights add to its charm and beauty. Built in the fifteenth century by Rana Kumbha, it is devoted to the Jain *tirthankara* Rishabdeoji with four sub-shrines, twenty-four pillared halls, eighty domes resting on four hundred pillars. The corridor has niches for *tirthankara* images, with each niche having a spire and little bells festooned atop each; so that the very breeze blowing in this temple seems to create celestial music.

The temple of Karni Mata located at Deshnoke, thirty kilometers from Bikaner, is dedicated to the early fifteenth century mystic-incarnate of the brown rats. They scurry around the main deity of the temple and these rats

The greatest polo teams were raised in Rajasthan where the princely kingdoms found it an ideal sport. While different versions of it—elephant and cycle polo, for example—have been played elsewhere in the state, in the desert, camel polo too is a whimsical adaptation.

are supposed to be repositories of the souls of dead *Charans*, the traditional bards. Milk, grains and sweets are offered to them during prayers each day. Rare white mice are especially revered. This temple is the main centre for the Navratri (nine-day Hindu festival) fair in September-October every year and thousands of devotees come to offer prayers. About fifty kilometers from Bikaner is the township of Kolayatji with a temple dedicated to the saint Kapil Muni and this is also an important site for annual fairs in October-November, the Indian month of Kartik.

Among other temples of significance are the Shri Ramdev temple at Ramdeva Ruincha which has a shrine of an incarnation of Ramdev. Besides being a place of pilgrimage, it has also become a centre for colourful cultural events.

Indeed, the desertfolks' love for colour seems almost a reaction to the desolate, dreary environment. Once the Rajasthani learned to survive in the desert, he moved a step beyond and learnt to live and produced a rich culture in the form of art, music, folk dances, various types of attire, fairs and festivals.

The colourful *pagdi* (turban) is one of the most important parts of Rajasthani attire and its bright colours make a striking sight against the golden sands. The number of strips tied around a *pagdi* decides the status of a person, the colour signifies the occasion and the style shows the caste and clan. It is necessary to wear a *pagdi* not only to receive respect, but also to give respect to others. A man cannot go before a lady without a *pagdi*; even members of the royal families must wear it before going to the chambers of the ladies. If a visitor wearing a *pagdi* is not properly welcomed, it is considered an insult, often resulting in family feuds. In fact, even to see someone bare-headed when going out on work is considered inauspicious. On the death of the father, the elder son wears a white *pagdi* and automatically gains control over all his father's property.

There's a saying in the Thar, 'those awake have run away with the *pagdi* of the one who has fallen asleep'—stealing someone's *pagdi* is to defeat and humiliate him. In the Thar every princely state has its own symbolic *pech* and *pagdi* according to individual customs and traditions. The Rathores of Marwar were renowned for their flashy turbans. Similarly, moustaches vary with caste, clan, status and upbringing in this macho society. They are symbols of valour, prestige and honour. The Rajputs attach great importance to them.

The people of the Thar have a festival for every season and these are enthusiastically observed. In fact, fairs and festivals are the backbone of the desert society where nature and the elements are held in deep-seated reverence. Gangor (in April) and the festival of Teej (July), both symbolise a reverence for scarce water. With the advent of the monsoon and the onset of the first shower, Teej is observed with a colourful procession in most towns and the erstwhile state capitals. It is headed by an elaborately decorated and richly painted group of elephants followed by women carrying earthen pots on their heads.

There are some festivals which have been specially designed to attract tourists. The Desert Festival held in Jaisalmer in February and the Bikaner Festival in January are examples of such events recently established by the Department of Tourism, Government of Rajasthan.

The Jaisalmer Festival (February 9-11) on the elevated sand contours is a landmark; the desert fort town of Jaisalmer attracts almost twenty thousand to twenty five thousand tourists from home and abroad. The whole place

Opposite page: The shrine of Deshnoke, close to Bikaner, is unique because rats are allowed a free run of the sanctuary. Built in honour of Karni Mata, this marble temple with silver doors enjoys a popular following in the region.

becomes vibrant and fraught with colour, music and dance. Exotic colours and intricate designs have always been the mark of desert culture and festivity. *Ghagras, burkhas* for camels, traditional elegant turbans, graceful well-kept beards and whiskers of menfolk like Karna Bheel (who finds mention in the Guinness Book of Records for his 7.10 feet long moustache), *dhoti* (unstitched length of cloth worn by Indian men) and *kurtas* (long shirts). Men wearing earrings, necklaces, silver lockets, colourful richly designed Rajasthani *jutis* (shoes) along with many ivory *chudas* (bangles) that women wear; these are all part of the festivity and enjoyment.

Though gold is worn by the Rajput and Marwari inhabitants of the desert, the tribals tend to prefer silver, and that too in huge quantities. They also go about their household chores bowed down under the weight of their ornaments.

Opposite page: *The love for dazzling colours and jewellery is manifest in the clothes and ornaments of the villagers who inhabit the desert. The Gujar community, in particular, wears heavily embroidered skirts and blouses with heavy pieces of jewellery.*
Following pages 84-85: *The Mangas and Langaniyas of Jaisalmer provide the music for the desert communities, singing ballads and also acting as minstrels as they lend melody to the exploits of family ancestors, often contemporising tales of greatness.*
Pages 86-87: *Women tend to gather together, away from the male gaze, as they sit heavily swathed to watch some popular entertainment. Their skirts and veils provide them protection both from the extreme heat and cold of the desert climate.*

Women walk in procession with silver *kales* (pots) on their heads which is a traditional symbol of an auspicious beginning. There are various mock competitions that add colour to the festivities. In the turban tying competition, the enthusiasm among foreigners is remarkable as they take active part. As also in the eating competition of *ghotasva laddu* to the accompaniment of music. Artists from Jodhpur, Barmer and Kota present music and folk dances like *Kalbetia, Teratali, Chari* and the popular *Ghoomar*. The whole golden city and the fort of Jaisalmer is quickened to life during this festival.

A group of musicians entertain villagers with their stringed instruments which they are extremely adept at playing. All festivities, marriages and celebrations begin with a musical prelude.

Other remarkable fairs and festivals are Kanana Fair (March), Barmer, Marwar Fair (February), Jodhpur and Nagaur fairs. The sleepy towns of Barmer and Nagaur are suddenly shaken out of their slumber. Enchantment and excitement prevail and are expressed through colours and festivities. They present inspiring spectacles of Rajasthani folk music and dance. Spirited folk ensembles perform with gusto in the Mandore gardens, a place about eight kilometers away from the city of Jodhpur. The Kanana fair symbolises the harvest period in the spring season. *Gair* dancers in colourful traditional attire assemble and perform dances for entire nights.

The folk music and dances of the Thar are remarkably alluring. Even the musical instruments, though they look simple, are acoustically sound and attuned to the classical *ragas* (musical melodies) and *sur* (notes). The richness and diversity of the folk music is part of a well-kept and continuing heritage, with a tradition of feudal patronage and interaction with neighbouring cultures. Scores of fairs and festivals open out a quaint mixture of *ragas* and moods where men and women take part with equal fervour and animation.

Devotional songs are part of the folk music and are enriched by the verses of saints like Kabir, Meera and Malookdas. They are spontaneously performed

The Kalbeliyas are from a community of snake charmers whose womenfolk perform a mesmerising dance using body movements that are extremely rhythmic. They are often seen in concert at important cattle or camel fairs.

during great festival days, when *raatijagas* or *jaagarans* (overnight singing) are held as part of the festivity with devotion. However, there are some songs which are timeless favourites with people and are always sung during festivals, like *Panihari, Ghoomar, Kunjan* and *Gorbandh.* Besides music, there is *Taalbandi* unique to the group singing of classical *bandish* (composition) specific to a region.

Puppetry is another art in the Thar which combines folk music and the art of story-telling. The themes are taken from Rajasthani folk tales, mythology and romantic ballads. Though popular in the whole of the Thar, Shekhavati and Marwar have carved a special niche for themselves, patronised by the nobles as well as the common folk.

The feudal set-up of the medieval age extended patronage to the musicians of the region and gave birth to a large number of hereditary professional singing communities. This impetus given to music reaped rich rewards as it evolved and became more sophisticated. The music varies according to the occasion, theme, the quality of the performer and his style. This brings in enough regional variation. For example, the Bhopas recite ballads and long narratives about Pabuji, Devji and render episodes from epics like the *Mahabharata* and the *Phad* of Pabuji in a traditional style.

The musical instruments are broadly of two types: chordophonic and aerophonic. In the first group, the Sarangi, a string instrument, has all along been considered an important folk instrument. Among the aerophonic group are flute-like wind instruments like Peli, Algoza, Satara, Narh or Nad, Pawri (a flute of the Kathodia community), Nafeeri and Surnai. There are also rudimentary forms of the Shehnai to be found in the Thar.

There are two more minor groups of musical instruments. The antophonic include Ghanti or Ghanta, Ghunghroos and different types of Manjeeras with their variants like Jhanit and Taala. Then there are membranophonic instruments like Damru, Dugdugi, Dhol or Dholi, Daf, Chang Dhaphli and Nagaras.

Among the folk dances, *Ghoomar* and *Gair* are very popular. *Ghoomar* is a dance where women swirl in a circle to a *kaharwa* beat and is ceremoniously performed during the Gangor festival. The *Gair* dance is performed by men before and after Holi. In parts of the Aravalli region it is a special favourite and occupies a prominent position. The Shekhavati area has the *Daf* dance in which male performers, some in female attire, alternately sing and play the *Daf*. During Holi, *Daf* is also associated with *Dhamals,* which are Holi songs. Then there is the *Bam Rasiya* dance in adjoining areas of Brij and also the *Chari* dance where women dancers balance many earthern pots on their heads while dancing. The *Bam Rasiya* is a zestful and uninhibited dance associated with Holi, performed by rows of swaying performers.

The camel has been deservedly celebrated in song and legend. After all, it is such an integral part of the landscape that it is difficult to think of the Thar without the large herds of camels at water-holes and the caravans of camel riders who fill the cool desert nights with sweet melodies. In Thar folklore, a large number of communities like Bhopas, Pabuji, Dharhis, Dholis, Kalbeliyas, Mirasis and Jogis sing songs to their beasts of burden, best friend and faithful companion.

Love legends occur almost in every part of the country and they are immortalised in folk music, like the love story of Moomal and Mahendra which has been immortalised in the sandy tracts of Jaisalmer. The ruins of Moomal-ki-Meri still exist in Lodurva near Jaisalmer.

Sam sand dunes outside Jaisalmer are celebrated for folk music performances at sunset when travellers, tourists and connoisseurs rush to the golden edge of the sand dunes and gather to listen to the folk music programmes, performed mostly by Manganiyar artists. The wrinkled and silver-bearded Chimme Khan, is a master artiste who's golden voice is well-known. It is a moment of sublime beauty to listen to the old folk songs handed down through the centuries as the golden sun sets on the desert making the sands shimmer.

This spirit of celebration is amply reflected in other ways too. The people of the Thar live in a dry and desolate land and yet have developed an overwhelming interest and desire to surround themselves with colour. Colours, in fact, have become the great leveller in this society. Even the migratory

Gair dancers from Jaisalmer perform at one of the several melas *or fairs in the desert. These events provide the occasions to combine commercial activities with pilgrimages and social interaction.*

gypsy community's womenfolk wear richly ornamented and profusely coloured clothes. The sight of these wonderfully colourful men and women against the stark backdrop of the Thar is a truly dazzling one.

CONTINUITY AND CHANGE

The Thar in recent years has undergone a complete metamorphosis as a result of the on-going process of development and change. Between 1891 and 1991, the population has grown tremendously, especially in the last few decades. Life has become a lot more comfortable, facilitated as it is by scientific and technological advances in water management, ecology, agriculture,

A crop of chillies dries under the hot desert sun. These searing red chillies will be ground and used to add spice to the cuisine of the local population of the Thar.

industrial development and transport. Water development projects like the Gang Canal, Indira Gandhi Canal and Luni Basin have brought many new areas of the desert under irrigation. The Central Arid Zone Research Institute (CAZRI) of the Government of India, and the Desert Development Department of the Government of Rajasthan have worked toward creating a better understanding of desert vegetation. Vegetation regeneration programmes are being undertaken on a war footing, to tackle overgrazing, over-cultivation and exploitation of resources, which otherwise lead to desertification.

The latest effort to revamp the economy of the Thar is through tourism.

Innovative planning based on the geography and culture of the region has made Rajasthan the foremost tourist destination of India. The region's beautiful palaces, impressive forts and exquisite temples draw thousands of visitors annually.

The royal museums of Jaisalmer, Jodhpur, Bikaner and other places in the Shekhavati region also draw a large number of tourists. The museum of Jaisalmer was opened in 1984; Marwar's magical past has been painstakingly re-structured in the Jodhpur museum. The Sardar Government Museum started way back in 1936 at Umaid public gardens attracts a large number of tourists.

A group of traders wind up a desert camp. For most, the camel provides the transport, and the sand the bed. Twigs and cowdung patties are used to light fires over which meals are cooked.

It has three main sections, and twenty-three sub-sections, featuring pottery, antique silver and gold, guns, and the like. Of particular note is the sculpture section exhibiting pillars from the 5th century AD. The rich Mehrangarh Fort museum documents the valour and bravery of the Rathores.

Fairs and festivals like the Nagaur and Marwar fairs and the Desert Festival mentioned earlier are already becoming major tourist attractions. The

Overleaf: *Out alone in the desert, the man and his beast are comrades-in-arms for they must eke out a life under some of the most savage conditions, but a life full of colour and celebration.*

development of tourism has created a number of wayside inns and guest houses. The most admirable are the individual efforts to set up heritage hotels like Hotel Castle Mandawa, Khimsar Heritage hotel and Dahu-ri-Dhani, in which the age-old architectural elements of these erstwhile castles are maintained but the interiors have been modernised to provide facilities for tourists and visitors. It is a common sight at these palaces to find the owners narrating tales of the fort or explaining the palace's ancestral bearings, its fortunes and disasters to wide-eyed, wonder-struck tourists.

In order to optimise tourism in the region, the state government is also trying to promote adventure tourism. The Thar's physiography obviously does not allow for the usual mountain safaris and adventures but keeping the desert topography in mind, the authorities have come up with innovative concepts like camel and canal safaris, cycle safaris and vintage car safaris. These take one through the golden sands of the Thar, the untamed rural landscape, the assembly of folk musicians and dancers—and the exploration of Rajasthani cuisine, of course.

The best camel safaris are in Shekhavati, Bikaner, Jaisalmer and the Fatehgarh *tehsil* of Jodhpur. Chhatargarh and Mohangarh provide great opportunities for canal safari. Special safaris are organised during the fairs and festivals of Marwar, Nagaur and Jaisalmer. Camel and jeep safaris in the Sam *tehsil* of Jaisalmer are special attractions, as nothing compares with the experience of listening to the simple but soul-stirring music of the Rajasthanis as the sun descends into the distant sand dunes.

As can be imagined, the cuisine of the Thar is very much a product of its geographical, cultural and historical background. The stress is more on nutrition, 'filling one up', rather than fuss and ostentation. Since the main meal comprises the few basic dishes, the Rajasthanis tend to 'spread' themselves a bit when it comes to sweets. *Lapsi* (sweet porridge), *churma* (wheat bread fried with ghee and sugar), *malpuas* (pancake and syrup) and *laddus* (small golden rounds) are popular sweets. Some of these have become Thar 'specialities', and the humble *dal bati* (flour dumplings in lentil curry), *churma*, *rabdi* (thickened milk) and so on served in gold or silver *thals* (plates) in restaurants are renowned delicacies.

All this has helped generate extra income for the people and has given a big boost to local art and specialised cottage industries. Indeed, tourism holds the key to a golden future for the Thar. The desert environment is dry, dreary and desolate but out of this drabness has emerged one of the most vibrant folk cultures, one that abounds in colour. There is so much zest for life in the desert folk that they never give up trying to outwit the uncertain and inclemental climate. Today one notices a renewed vigour and effort to turn the desert into a land as golden as the sands it is famous for.